A Modern Introduction To
HEMP
From Food To Fibre

The healing powers of hemp are truly broad spectrum: as a source
for fibre for textiles, automotive composites, building materials and
plastics; hurd used as an industrial filler and for animal bedding; seed
as a nutritional grain and a source for oil for biofuel, cosmetic use and
nutritional supplement. Hemp as a crop needs no pesticides unlike
cotton, improves soil structure, can substitute the forests as a supply for
paper and can be grown throughout the world.

**Dedicated to sustaining life, alleviating
suffering and bringing happiness to all beings**

Welcome,

To a modern introduction to the mighty hemp plant, and ways it may benefit our society. I have incorporated new growing technologies that are playing a significant part in the positive changes this world needs, socially, commercially and environmentally.

You will find facts on the history of hemp around the world. My passion in regards to hemp foods has led to a more indepth look into the nutritional profile of hemp.

The hemp industry is also moulding its way into the automotive, textile, cosmetics and plastics industries. And this is just the start.

I vision hemp based materials in nearly every type of product. This can only be supported by our choices as consumers. We must support the networks of companies pushing boundaries with newer, higher quality and value for money hemp products. This is why I have accepted sponsorship at the end of this book- to support this sprouting industry.

Thank-you for supporting industrial hemp.

Paul

What Interests You About Hemp?

✓ crop grows in less than 100 days

✓ improves the condition of the soil

✓ great fibre for clothing and textiles

✓ sustainable source for paper production

✓ stronger and more durable than alternative fibres

✓ alternative to petrochemical based plastics

✓ environmentally friendly building material

✓ hurds absorb five times their own weight (great oil absorbent)

✓ seeds are 25% quality protein

✓ oil is high in omega-3, 6 and 9

✓ source of fuel for diesel engines

✓ medicinal hemp noted as a pain reliever

✓ bringing cash back to the farmer and community business'

✓ long proven history

CONTENTS

A Modern Introduction To

HEMP

From Food To Fibre

A CROP FOR FARMERS
A FIBRE FOR MANUFACTURERS
A SEED FOR US ALL

What is Hemp?

Hemp is a variety of *Cannabis sativa L.* that is a tall, vigorous annual plant that often reaches a height of over four meters at maturity. *Cannabis sativa* shares the Cannabaceae family with only one other genus, Humulus, which includes hops. Hemp and marijuana are both varieties of *Cannabis*, but industrial hemp contains very little (0.3 percent, or less) of delta-9-tetrahydrocannabinol (THC), a compound that produces a psychoactive effect, while marijuana contains much higher levels of THC (up to 30 percent). If you smoked a field of industrial hemp, you would only get a headache.

Only recently has hemp seed or grain been accepted by mainstream organizations as a safe and natural food. Consumers appreciate hemp grain for its health benefits and pleasant nutty flavor. The seed provides essential nutrients, including high quality protein, soluble and insoluble fibre, vitamins, minerals and essential fatty-acids (EFAs). In fact, hemp is the only natural food that contains both EFAs in a balance ideal for human consumption.

Not only are hemp seeds nutritious, the plant can also be made into rope, clothing, body care products, and medicine, as well as a base for plastics, paper, and fuel. Hemp produces four times as much paper pulp over the same period of time as the equivalent area of trees. Since trees take up to 30 years to mature, and hemp grows in 90 days, it is astonishing that so much virgin forest is still used for paper.

Hemp History

Throughout recorded history, the people of Asia, Africa, and parts of Europe have used hemp for fibre and enjoyed the rich nutritional value of hemp seed foods. The Greek historian, Herodotus, recorded that hemp was used in the manufacture of linen in Scythia. The oldest Chinese agricultural treatise, the Xia Xiao Zheng, written in the 16th century BC, lists hemp as one of the major crops grown in ancient China.
It was not until 1937 that a corporation called Dupont Industries became involved in the hemp chronicle. Dupont, and other influential parties, including newspaper magnate William Randolph Hearst, had significant interest in the synthetic fibre,  logging and paper industries. A campaign was launched to dirty the name of hemp, as it was potential competition for the new industrial materials being created. Unsupported claims ("the killer weed") and a term

Hemp Tradition

It has been suggested that the formerly unidentified Hebrew word, Tzli'q, (Tzaddi, Lamed, Yod, Quoph), refers to a Jewish meal of roasted hemp seeds that remained popular in medieval times and was sold by Jews in European markets.

In Europe, it was once required that monks eat three meals made of hemp seed daily, whether in soup, gruel, or porridge.

Hemp seed oil is said to burn the brightest of all lamp oils and has been used since the days of Abraham. Scythians once used hemp oil to purify and cleanse themselves.

In Latvia, hemp seed is traditionally included in festival foods eaten on St John's Day, June 21st.

Polish cooks bake hemp seeds into holiday sweets.

imported from Mexico, marijuana, eventually led to most of the western world banning a crop that once had a bright future.

In a major turnaround not so many years later, we found the governments of our nation demanding the growth of hemp and promoting its use in World War II using promotions such as the Hemp For Victory video produced by the US Government. Farmers were actually fined if they did not grow some hemp on their land! After World War II, hemp was again outlawed and large corporations grew to become even larger and more powerful multi-nationals.

Growing Hemp

GROWING HEMP IS A FAMILY BUSINESS

The sun allows a small sprout to grow quickly into a tall hardy plant. In the Northern Hemisphere, seed is sown at the end of March and harvested for fibre at the end of August with seed being harvested in September. The date varies depending on the soil, variety and latitude. The most important factor for seed production is a dry harvest and ability to achieve an even-ripening of the seeds.

Hemp is a wonderful rotation crop. It has the ability to grow below

PHOTO: EIL

HEMP WORKS BEST IN CROP ROTATION

the soil nearly as fast as it grows above. A fully mature hemp fibre crop can reach up to five metres. The tap-root can be over one metre, so the plant is able to draw nutrients from deep in the soil. Few crops are that fast growing. Hemp's distinctive leaves assist this growth. As the plant grows, its leaves are continually shed, ensuring that most nutrients are recycled into the topsoil by harvest time. The speed at which densely sown hemp forms a leaf canopy, coupled with its overall speed of growth, means that very few weeds can compete, making herbicide unnecessary. Hemp also has its own built-in insecticide called terpene. The terpene content is responsible for hemp's distinctive pungent smell, and is believed to repel insects. Hemp leaves have little commercial use, except as part of the production of essential oils. However, they do have the potential to feed us indirectly. Tests have shown that hemp fodder, used instead of alfalfa meal, provides the same rate of growth in chickens and the same or even better rates of egg laying. Eggs from chickens fed on a hemp diet have intensely yellow yolks due to the high levels of carotenoids in the hemp seed. Carotenoids are fat soluble pigments said to provide health benefits in decreasing the risk of disease, particularly certain cancers and eye disease. The by-product of hemp seed oil extraction, hemp seed cake, is also an excellent animal feed (see pg.15)

Hemp in History

The first known fabric, from 8,000-7,000 BC, was apparently woven from hemp fibres.

The legendary Chinese emperor and physician, Shen Nung who lived around 2300 BC made the first recorded references to *Cannabis*. He recommended doctors use the hemp superior elixir of immortality to treat patients with everything from constipation and menstrual cramps to depression and gout.

Archaeological evidence proves hemp has been cultivated since prehistoric times, with the earliest findings at a 12,000 year old Neolithic site at Yuan-shan in Taiwan

Hemp is mentioned in Buddhist doctrines dated ca. 500BC

In 16th century BC, the Egyptian Ebers Papyrus recorded the medicinal use of cannabis.

Hemp Varieties

Hemp is grown all over the world. Different varieties thrive in local conditions. Historically, high latitude temperate climates produce hemp seeds with higher quantities of omega-3, while fibre strains are better suited to tropical regions. Most fibre varieties are tall reaching three to five metres under optimal conditions. In the initial stages of industrial hemp production, these heights presented some difficulties for modern farm equipment, though such problems have now been overcome.

> Hemp needs no pesticides because it is already unpalatable to most insects. Hemp needs no herbicides because, under the proper conditions, it grows too quickly for any weeds to compete.

A variety has been developed & trialled by Ecofibre Industries in Australia that is suitable for temperate regions, but thrives in sub-tropical and tropical climates, grows to five meters and produces 18.5 tonnes of biomass in just 90 days.

There is an ever-expanding variety of industrial hemp strains, all of which are being bred without genetic modification, to suit a range of environments and uses. One variety of hemp, originally from Mao Luo Island of China, in the Eastern Sea, was said to produce seeds of the "highest quality" that grew to the size of garden peas!

Hemp in History

The Vikings used hemp rigging on their infamous voyages.

In 6th century Persia, a preparation called Sahdanag (meaning royal grain) was made from hemp seeds for use in the spiritual matters of royalty.

In 1175AD hemp is mentioned in the seventeenth book of The Holy Grail of King Arthur.

In the 15th century, the Gutenberg Bible was printed on hemp.

16th century England relied heavily on hemp cultivation and importation. During the Tudor period, the British navy was so dependant on hemp for rope and canvas that farmers were fined if they did not grow it.

In England in the 18th century, full British citizenship was bestowed by a decree of the Crown on foreigners who would grow *Cannabis*; fines were often levied against those who refused.

High quality, food-grade hemp seed needs a balance of omega-3 and omega-6 fatty-acids. One of the first superior varieties of hemp grown for food-grade seed contains double the quantity of gamma-linolenic acid (GLA) and steridonic acid (SDA) of currently existing hemp. This variety is known as Finola™. People in Central Russia and Asia have developed shorter, earlier blooming varieties, through hundreds of years of selective breeding. Finola™ was developed from this stock and reaches a maximum height of one and a half metres and does not tend to branch, thus producing a single stalk that is typically full of seeds. This variety also looks less like typical *Cannabis*, which reduces the need for intensive policing of this variety, compared to most conventional hemp crops.

The primary advantage of a short variety of hemp is that it can be harvested by a conventional grain harvester or combine. The crop may be swathed or straight-combined at low moisture, without the danger of plugging up the machinery. Over two metric tons of seed per hectare have been produced from Finola™ which typically produces more oil (up to 38%

Hemp in History

The war of 1812 involving America and Great Britain was sparked by the desire of each country to dominate Russian hemp supplies. Russian hemp production was also the primary reason Napoleon invaded Russia in 1812.

The American Declaration of Independence was written on hemp paper.

George Washington and Thomas Jefferson grew *Cannabis* on their plantations. Jefferson, while envoy to France, went to great expense and even considerable risk to himself and his agents to procure particularly good hemp seeds smuggled illegally into Turkey from China. The Chinese Mandarins (political rulers) so valued their hemp that they made exporting the seeds a capital offence.

Even ex-President Clinton indulges in products made from this ancient grain.

more) and protein (up to 28 percent more) than other hemp varieties. The fatty-acid profile shows a content of over 90 percent unsaturated fats, with over four percent GLA and two percent SDA, making it a superior variety for human consumption. The seeds are slightly smaller and sweeter, with up to five percent more natural sugars, than conventional hemp seeds. The seeds also have a delicious citrus-pepper taste.

HEMP BAST AND HURD

65 % Bast Fibre: Long, slender fibres from outer stalk

35 % Hurd: Core fibre from the center stalk, high in cellulose

containing:
71 % Cellulose and <25 % Natural Lignin

Hemp hurd, water and lime are used in France to produce a natural mould and insect resistant building material that is stronger than cement- at 1/6th the weight

Hemp in History

Queen Victoria's physicians prescribed cannabis as a treatment for menstrual cramps.

In Australia, hemp seed was used during two famines in the 19th century.

Adidas and Converse have manufactured hemp sneakers, Calvin Klein has offered hemp bed spreads and Georgio Armani has developed high fashion hemp clothing.

Hemp can grow in different climatic zones throughout the world and is an economically viable and ecologically sustainable alternative, or compliment to conventional crops.

FINOLA CROP

PHOTO: EIL

Hemp Around the World

In Germany, where cultivation of hemp has only been reintroduced since 1996, the first year's harvest covered approximately 1422 hectares. In 1998 this had more than doubled to 3500 hectares. Two years later the area had more than doubled again, and Germany is still only a minor player in the European field of hemp farming. In 1999 France grew 11,000 hectares, and Spain grew even more. Canada grows more than 20 varieties of hemp, in excess of 14,000 hectares. In 2002, Australia accepted open growing of industrial hemp in the state of Queensland. Down under they now forecast the potential to grow 50,000 hectares of hemp within four years, a remarkable level of growth by any standards. Australia and other parts of the world see the potential for this truly versatile crop, and the jobs and economic benefits that the hemp industry will bring.

As demand increases for hemp products, more processing facilities are being built around the world. One new industrial approach to hemp and flax fibre extraction is enzyme fibre extraction. Another is known as total fibre processing. The aim is to develop low cost, pollution free, fast and efficient cultivation and processing technologies to enable flax and hemp fibres to overtake other fibre crops in modern industrial markets. These are small to medium sized on-farm operations that are contributing to local sustainable non-polluted environments. Hemp farming co-operatives are being set-up worldwide with a unique system of fibre processing. An example of this is where ten farmers each grow between 50 and 500 hectares that are processed at such an agreeably placed mini-processing facility.

The farmers support the processing facility, which supports them by upwardly valuing the crop, enticing farmers with a higher return than conventional crops.
Such localised machines are often suited to one particular industry. They may be designed to produce long clean fibres for the textile industry, short fibres for pulp, paper or plastic industry, or another quality for non-woven materials. Some co-operatives have sensibly invested in harvesting machinery specifically aimed at cultivation of the seed. This works in conjunction with a seed processing plant that is suitable for commercial oil production or seed de-hulling. The co-operatives will usually include members other than farmers, including groups that can market the product and possibly increase the value of the crop in some other value-added form.

International Terms for Hemp
Cañamo — Spanish
Chanvre — French
Konoplya — Russian
Kender — Hungarian
Da Ma — Chinese
Hennep — Dutch
Hanf — German
Hampu — Finnish
Kannab — Persian
Kanop — Armenian
Kanas — ancient Celtic
Konopj — Polish
Hampa — Swedish and Danish

Hemp and Flax

Hemp and flax crops are used for both their fibres and seeds. Both flax and hemp fibres are viable alternatives to more conventional fibre crops. The development of the hemp and flax industry in Europe over the last ten years is directly related to a growing

SEED YIELD COMPARISON

Hemp Seed	Flax Seed
approximately 900kg/ha	approximately 800 kg/ha
yields of up 2000kg/ha have been recorded.	yields vary from 500 to 2500kg/ha

market for plant fibres. As these old crops are rediscovered by industry, economical and environmentally sustainable processing technologies are being developed. New machinery has increased the viability of long and short fibre crops. Recent increase in demand for seed makes dual use for a possible single crop, allowing less waste and higher returns.

The UK is currently the biggest consumer and importer of flax worldwide and is quickly becoming interested in the alternative, hemp, for its greater yield and beneficial nutritional properties.

PHOTO: EIL

AUSTRALIAN VARIETY TRIALS

Characteristics of hemp and flax (linseed)		
	Flax	Hemp
Name	Linum usitatissimum	Cannabis sativa
Climatic regions	Temperate and sub-tropical	Temperate, sub-tropical + tropical
Fibre length (ultimates) mm	20 23	9 20
Fibre diameter (µm)	22	22
Cellulose (%)	56 64	67
Straw yield (tonnes/ha)	2.5-7	5.0 7.0
Fibre yield	20%	20%

Edible Hemp Products

Why Hemp Foods?

✓ a great source of quality protein

✓ perfect balance of essential fatty acids

✓ contains good quantities of soluble and insoluble fiber

✓ replacement to soy products- safety of GM contamination

Functions of Essential Fatty Acids in Hemp Seed

✓ improve energy levels, stamina and recovery from fatigue

✓ make skin soft and velvety and help with acne, psoriasis and eczema

✓ help weight loss by increasing metabolic rate

✓ help in healing of wounds and injuries

✓ improve the symptoms of ADHD

✓ improve mineral transport and mineral metabolism

✓ transport cholesterol, lower high blood pressure and decrease risk of unwanted clots in arteries

✓ assist with premenstrual mood changes

✓ curb cravings for sugar, and junk food fats

✓ improve the functions of all the organs and glands

✓ assist in brain development and adult brain function

Hemp Seed

WHOLE HEMP SEED	The seed of the hemp plant is a small, soft oily nut, surrounded by a thin transparent inner layer and a hard shell. The hemp seed vaguely resembles the seed of other cultivated grains, such as wheat and rye. While raw whole seeds contain the highest percentage of fragile oils and beneficial nutrients, according to current law, in North America raw whole seeds sold must be	DEHULLED HEMP SEED
* crunchy * highly nutritious * may contain minor traces of THC * can be stored longer than dehulled seed * as nature intended		* improved texture * no grit between your teeth * more densely packed nutrients * will not germinate * contains only trace quantities of THC

sterilized or the hull removed to prevent sprouting. Thanks to recent processing technology, the husk can be removed to reveal the pure, nutritionally-dense hemp nut. The soft nut meat is the most useable part of the hemp seed and a versatile ingredient in many recipes. Store hemp nut in the refrigerator. Whole hemp seeds still have their place; when toasted they are a tasty, crunchy snack.

Toasted Hemp Seeds Recipe

1 cup hemp seeds
1 cup sunflower seeds
1/2 cup water
1/2 cup Tamari or Soy Sauce
pinch of cumin seeds, or to taste

* Soak hemp seeds and sunflower seeds for 1 hour in water and Tamari or Soy Sauce to cover.
* Drain and roast with cumin for 3-5 minutes in a dry pan, continuosly mixing.
* Do not let dry or change colour.

By keeping their moisture, seeds do not loose their life-force.
Great as a topping for salads or a snack to be enjoyed any time. Best for kids (like me!)
Try it with just Celtic sea-salt, sweeteners or herbs.
Re-use the water/ tamari mix to marinate vegetables (mushrooms are nice).

Hemp Oil

Hemp Oil is produced by cold-pressing fresh hemp seeds. Hemp oil is my preferred way of taking a nutritional dose of hemp; health professionals often recommend hemp oil over other oils for reducing symptoms of eczema, cardiovascular, and the menopause. I also recommend hemp oil as it tastes great.

To get the most out of hemp oil, ensure the product is fresh, stored in a cool (refrigerator) dark place (an opaque bottle is fine) and not heated (add at the end of any cooking) to ensure the fragile omega-3 and omega-6 content is not spoiled. Hemp oil best not used in cooking, but as an edible nutritional oil.

Production of hemp oil is usually carried out in a hydraulic screw press with a maximum heat of 45° C (110°F). Lower temperatures produce nutty, hazy oil. Higher temperatures produce darker, greener oil. More high-tech methods of producing oil include super critical fluid extraction (using carbon dioxide). Such methods are designed to efficiently extract the oil from the seed, without

> **The Seed**
> A seed is a library of information, a storehouse of energy and a potential life. The seed is an excellent vehicle. It can travel around the globe in wind, survive fire, frost and adverse weather conditions for many years; when the time is just right, life can begin again. We are all dependent on seeds. Without seed there would be no plants, no colour green and no life. Honored in many cultures, the seed is a symbol of life's regenerative powers and has been traded as a prized gift since bartering began.

> Through our own creations and ideas of self-importance a boundary has been created separating ourselves from nature. We talk about the environment and health of the nation as if we are separate from them. This boundary does not serve us well. The time is ripe to re-discover respect for the Earth and its connection to ourselves, and each other. It is time again to incorporate the seed into our daily diet and acknowledge the life giving energies that it provides for our health and well-being.

chemicals and ensuring a safe, low heat. Hexane extraction is used for non-food grade hemp oil, often used for paints.

Hemp oil is best when unrefined as processing diminishes the nutritional value by de-naturing the essential fatty acids (EFAs). The best test for the quality of any oil is the taste test. It is easy to differentiate between the rancid, scratchy taste of poor quality oil and the smooth, nutty taste of superior oil.

Hemp oil contains mostly the gamma form of vitamin E, a natural anti-oxidant that keeps the oil fresh. "Nature does provide", but not necessarily for mass production and transportation. Buy fresh, and when possible, locally cold-pressed oil. Better still, press it yourself from fresh whole or hemp nut. Keep hemp oil in a dark glass bottle, in the refrigerator.

Hemp Seed Cake

The by-product of oil pressing and the husks of hemp nut provide a product containing between 30 and 50 percent protein, as well as a good amount of quality oil. In fact, 35 percent of the available seed oil remains in the cake after oil pressing.

Hemp seed cake is sometimes added to beer to add a nutty, wholesome flavor.

The most popular use for hemp cake is as an animal feed. Hemp Cake is sought for horses, cats, dogs and birds as it improves most skin and hair conditions, amongst other reputed benefits. This is because the oil content still contains the EFAs and as the meal is so high in quality protein, hemp cake offers an exceptional option when compared with other oilseed cakes. It is best to mix this rich ingredient in proportion to the animal's particular nutritional needs. Before substituting regular animal feed with hemp cake it is important to seek veterinary advice.

Hemp Flour

Hemp flour is a high quality raw material available from primary producers and resellers. I make it from whole seeds, or hemp cake, at home, by using the fine setting of a coffee grinder. I soak the seed in advance of grinding (for approximately 24 hours) to increase the enzyme content, making the flour easier to digest, and producing a more dough-like texture. Hemp flour has a very short shelf-life*

To use hemp flour, replace 10 percent of the wheat flour with hemp flour in any bread recipe. The trick to making fresh bread is to knead well. Ten minutes of kneading really does build your muscles, however the longer you knead, the lighter the bread. Use fresh yeast and vary the mix of flours. Try spelt, whole-wheat and hemp. Adding the soft, whole hemp nuts makes a delicious, nutty bread.   *see shelf life guideline next page*

Hemp Protein

A high protein powder isolate has been developed to take advantage of the high quality of protein available from hemp seed. Protein powders may be used as supplements by those requiring extra quality proteins. Powders can be added to fruit juices or smoothies. Being low in fat and containing no carbohydrates, this product is more stable than any other hemp product. Nutritionally superior hemp protein powders are becoming available at a price competitive to other protein powders.

Hemp Essential Oil

The Swiss first developed "hemp essence." First thought to unavoidably contain large amounts of THC, research for mainstream use was abandoned. They were allowed only in specific medicinal trials. Today, essence of hemp is being developed without THC, ensuring it is a harmless addition to food products suitable for children.

The refreshing, uplifting aroma of the pure essence offers great potential as a food flavoring and natural additive. Used in tea, soft drinks, beers, candies and ice cream, the essential oil has a gentle, but distinct, piquancy.

Improved harvesting and processing techniques mean the price of the pure oil has recently been reduced. The potent oil can be micro-encapsulated and used in a variety of dried products, including freeze dried soups and snack bars.

The essence is also a valuable base for a variety of body care products, although due to a difference in cost, some well-known companies have ignored the use of this ingredient and opted for patchouli instead.

Shelf-Life - a guideline

The shelf life of a hemp product varies, depending on the exact method of production, packaging and storage.

Whole hemp seed: 24 months from date of harvesting if seeds are ripe, kept in a cool, dark place, preferably packaged in lined cardboard tubs or thick paper bags

Dehulled hemp nut: 12 months from date of hulling, if seeds are fresh and kept in a cool, dark place, in an air-tight container. Keeping hemp nuts, moisture free, in the freezer is best.

Toasted hemp seeds: Fresher is better, although roasted seeds will keep up to three months in the refrigerator.

Hemp flour: 3 months from manufacture, kept refrigerated.

Hemp seed oil: 9-12 months maximum, when hemp oil is cold-pressed, packed in an oxygen-free environment and stored in a dark bottle in the refrigerator. Although oxygen is a major cause of rancidity, by far the greatest destructive agent is light. Light enhances damaging free radical production in oils turning them rancid. You can usually recognise rancid oil by its strange flavour and the scratchy feeling it leaves at the back of the throat. Freezing the oil extends storage significantly.

Hemp ice-cream: 18 months for dairy free ice-cream made with fresh hemp milk, kept frozen at 10° F (24° C).

Hemp seed snack bars and long-life breads: Shelf life depends greatly on exact method of production and quality of raw ingredients used. Some products made by manufacturers unaware of hemp processing requirements may not last more than 6 months before rancidity takes hold, so buy from a reputable source. Good-quality products that have been cooked at low temperatures for short periods, and stored in a cool, dark place (printed film rather than see-through packaging) can be stored for up to 12 months, though 9 months is recommended.

Fresh hemp bread: 4-7 days for freshly baked bread made using significant quantities of hemp flour.

Hemp milk: 24-48 hours for freshly produced, untreated organic hemp milk. Best consumed within 12 hours.

Hemp protein: 6 months for 'sieved cake protein', 3 years + for true protein isolate (less than 1% fat)- check with the producer of your brand or contact HFIA (see back).

Other cooked hemp food products: less than 24 hours if cooked for over 30 minutes above 110° F (45° C), then refrigerated. Leftover food loses nutritional value through the oxidation process. Cooked foods loose vitamins and minerals through the heating process. Un-cooked (raw) foods are alive.

Hemp Nutrition

A nutritional analysis of hemp seed is impressive. Compared to flax and soy, hemp is more biologically compatible with the human body. Many people have trouble digesting soy products due to the oligosaccharide content, which can cause gas and stomach upsets. Ground flaxseed releases cyanide gas because of the presence of cyanogenic diglycosides. Hemp does not create these problems.

Although a valuable source of fibre, proteins and EFAs, hemp is by no means a complete food. There is nothing on this Earth that is, but when used as part of a holistic lifestyle, with a diet high in leafy green vegetables and fresh fruit, you will radiate good health.

> "I also recommend hemp... Actually, I think hemp is superior to flax, given its chemical composition. If you have a chance to try hemp oil, a forgotten newly discovered food; I think you will see why I am enthusiastic."
> Dr. Andrew Weil, M.D.

Oil Chart

Temperature	Unrefined Oil	Usage
	DO NOT HEAT	*high in super-polyunsaturates*
115ºF/ 46ºC	Hemp Oil	Smoothies
	Flax Oil	Salads
	Walnut Oil	Pure
	LOW HEAT OIL	*high in poly.-unsaturates*
212ºF/100ºC	*maybe* Hemp	Short baking
	Pumpkin Seed Oil	Light Sauteing
		Salad Dressings
	MEDIUM HEAT	*high in poly & mono-unsaturates*
325ºF/163ºC	Sesame Oil	Baking
	Extra-Virgin Olive Oil	Light Sauteing
		Salad Dressings
	HIGH HEAT	*high in saturates*
375ºF/190ºC	Coconut Oil/ Butter	Frying
	Ghee	Baking

NUTRITIONAL ANALYSIS

Typical Nutritional Analysis of Hemp Seeds

Protein	23.5%
Carbohydrates	35.8%
Moisture	5.7%
Ash	5.9%
Calories	503 per 100g
Dietary fibre	35.1%
	(3.0% soluble)
Fat	30%

Typical Nutritional Analysis of Hemp Cake/ Flour

Crude Protein	29.6%
Crude Fat	4.9%
Carbohydrate	57.5%
Crude Fibre	27.6%

Minerals:

Boron	138.2ppm
Copper	19.1ppm
Iron	254ppm
Manganese	95ppm
Nitrogen	4.74%
Zinc	88.2ppm

Hemp Seeds Typical Essential Fatty Acid Profile

Omega-3 (Alpha Linolenic)	20%
Omega-6 (Linoleic)	57%
Omega-9 (Oleic)	12%
Stearic	2%
Palmitic	6%
Carotene (Vit A)	16,800 IU/lb
Thiamine (B1)	.9mg/100g
Riboflavin (B2)	1.1mg/100g
Pyridoxine (B6)	.3mg/100g
Niacin (B3)	2.5mg/100g
Vitamin C	1.4mg/100g
Vitamin D	100 IU/100g
Vitamin E	3 mg/100g
Vitamin B1	0.9mg/100g
Vitamin B2	1.1mg/100g
Vitamin B3	2.5mg/100g
Vitamin B6	0.3mg/100g

Typical Fatty Acid Profile (major acids) of Hemp Oil

	In % of total fatty acids
Palmitic (C16:0)	7
Palmitoleic (C16;1)	0.2
Stearic (C18:0)	3
Arachidic acid (20:0)	2
Behenic acid (C22:0)	<0.2
Total Saturated Fatty Acids	11
Oleic acid (C18:1)	12
Linoleic (C18:2)	56
Alpha-linoleic acid (C18:3)	16
Gamma-linoleic (C18:3)omega-6	4
Steriodonic acid (18:4)	1
Total Unsaturated Fatty Acids	89

Vitamin E	150mg/100g
(mostly gamma-tocopherol)	16IU/100mg
(as alpha-tocopherol equivalents)	
Chlorophyl	50ppm

Protein

Hemp seed contains more that 22 percent protein, which is highly digestible due to its globulin form, as edestin and albumin. Edestin is a superior type of plant protein, similar to protein found in the human body, and perfectly suited to the body's cellular needs. Hemp contains the highest percentage of globulin protein found in any plant. It is important to note that protein can become denatured if heated above 115° C (239° F) for more than a few minutes, making it insoluble and less digestible. Edestin proteins are forerunners to hormones, haemoglobin (which transports oxygen and carbon dioxide in the blood), enzymes (which control many biochemical reactions), and antibodies (which fight off invading bacteria, viruses and toxins). Edestin also assists in suppressing symptoms of sickness and disease by increasing the body's own defense systems.

Hemp Protein Analysis (overview)	
Substance	**amount, per gram of protein**
Arginine	15.1 mg
Histidine	4.8 mg
Methionine	3.1 mg
Cysteine	1.3 mg
Lysine	4.5 mg
Threonine	2.9 mg
Serine	4.8 mg

Hemp seed contains good quantities of arginine and histidine; both are important for growth during childhood. Hemp protein also has the sulfur-containing amino acids methionine and cysteine, which are needed for proper enzyme formation, as well as relatively high levels of the branched-chain amino acids that are important for the metabolism of exercising muscle. Hemp seed contains both essential and other amino acids useful for childhood growth. In fact, hemp seed contains all eight essential amino acids, mostly in levels superior to those found in soy protein isolate or flax seed.

Fibre

Thirty-five percent of whole hemp seed is composed of dietary fibre, which is nearly ten percent higher than the fibre content of flax seed. Ten percent of the dietary fibre in hemp seed is the water soluble fibre that helps maintain blood glucose levels and lowers blood cholesterol levels, a benefit for diabetics. The remaining 90 percent, the insoluble portion, helps to prevent constipation by increasing fecal bulk and reducing bowel transit time. It has been shown that people who consume a high fibre diet lower their fat and blood cholesterol levels. High fibre diets, especially those of plant origin have also been associated with lowered risk of cancers of the breast, prostate and rectum. High intake of foods of plant origin (all of which contain some fibre) is associated with a reduced risk of heart disease such as angina, prevention of cancer and an increased life expectancy. This is believed to be because the fibre absorbs and eliminates toxins from the body, before they do any real damage. Another benefit of a high fibre diet, one that may assist with weight control, is the feeling of fullness that follows a fibre-rich meal.

MALE FLOWERS

DISTINCTIVE SHAPE
OF THE HEMP LEAF

FEMALE FLOWERS

Fats

Fats are not all bad. Many people still believe that all fat from food leads to body fat, but it is a fact that consumption of essential fatty-acids (EFAs) can actually lead to **weight loss**. Eighty percent of hemp's fat consists of EFAs and a high percentage of EFAs in the diet is desirable.

Many 'low fat' and 'high in poly-unsaturated fat' products are sometimes more damaging to our bodies than regular and full-fat products, such as full-fat milk. Such low fat 'diet products' have often been hydrogenated, which effectively destroys the oil and renders the nutritional components as nutritionally deficient and often toxic to our bodies. Hydrogenation is a process of highly heating fats.heating fats in this way destroys all the good oils and leaves just the long shelf-life saturated oils - great for manufacturers profits, not so great for health. In contrast, high quality, un-refined hemp seed oil, can provide a near perfect mix of EFAs and are never hydrogenated.

> Hemp seed contains the most naturally perfect balance of omega-6 linoleic acid (LA) and omega-3 alpha-linolenic acid (LNA) for humans. These are two Essential Fatty Acids (EFAs) that humans require. They are essential because they are needed by the body but cannot be produced by the body; they have to be ingested via food such as hemp seed based products.

Alternative sources of omega-6 Linoleic Acid (LA) and omega-3 Linolenic Acid (LNA) include flax seed, flaxseed oil, evening primrose oil, cod liver oil and green leafy vegetables. Only hemp is found in the ideal ratio of 3:1 (omega-6 :omega-3). The Standard American Diet (SAD) is often skewered too much in favor of omega-6. We can use high omega-3 oils such as flaxseed oil to improve the balance, but only hemp has the

EFAs in the correct long-term ratio by itself.

> "One can develop omega 6 deficiency by using only flax oil for too long. Hemp seed oil can be used over the long term to maintain a healthy EFA balance without leading to either EFA deficiency or imbalance." —Dr. Udo Erasmus.

Hemp seed usually contains between two to four percent omega-6 gamma-linoleic acid (GLA). GLA is also found in other more expensive oils such as evening primrose oil and borage oil, but hemp oil is noted as being the best tasting. The benefits attributed to GLA are many; they include relief of premenstrual syndrome (PMS has been shown to be related to an obstacle in the metabolism of LA to GLA), and use in chronic skin diseases such as eczema and neurodermatitis. Due to the high levels of quality EFAs in hemp seed, it is imperative you also include plenty of green leafy vegetables and fruits to supply the body with vitamins and minerals that assist the conversion of EFAs into prostaglandins that decrease inflammation, water retention and blood pressure. Required nutrients include vitamins B1 (e.g., found in wheatgrass), B3 (e.g., found in avocados), B6 (e.g., found in bananas / avocados), C (e.g., found in kale / broccoli), D (e.g., synthesized using sunshine) and the minerals magnesium (e.g., found in tomatoes/ spinach), manganese (e.g., found in hemp / parsley / beetroot) and zinc (e.g., found in hemp /oysters / cashews).

> Hemp foods have a high content of antioxidants (92.1mg/100g) in the form of alpha-, beta-, gamma-, and delta-tocopherol and alpha tocotrienol.

The Usefulness of EFAs

So omega-3 and omega-6 are essential to health. But why? Omega-3 (LNA) has been shown to protect against certain types of cancers and positively modify immune and inflammatory reactions. It has been demonstrated that renal, respiratory, cardiovascular, and dermatological conditions are improved by including omega-3 in the diet. Omega-6 (LA) can help protect against acne, loss of hair, poor blood circulation, and cardiovascular disease as well as liver, kidney and gallbladder problems. EFAs are also necessary for maintaining the structure of cell membranes and the permeability of the skin. Health Canada recommends that pregnant and lactating women increase their omega-3 intake as found in the equivalent of approximately 1 tablespoon of hemp oil per day. EFAs actually assist in the most basic of bodily operations, such as helping with the transfer of bioelectric currents from cell to cell and contributing to brain function and development. EFAs convert lactic acid (often responsible for aching muscles during heavy exercise) to harmless water and carbon dioxide.

Harry Barnes, an international acclaimed triathlon sportsman recently arrived back from a tournament in Cancun where temperatures reached 102°F in the shade. For Harry 'it was a walk in the park'. At this tournament he placed third place, just two seconds from second place. "I will get better at this sport. I definitely have great faith in the hemp products that played a big part in my being able to train day in day out with great recoveries and being able to push my body to its full potential."

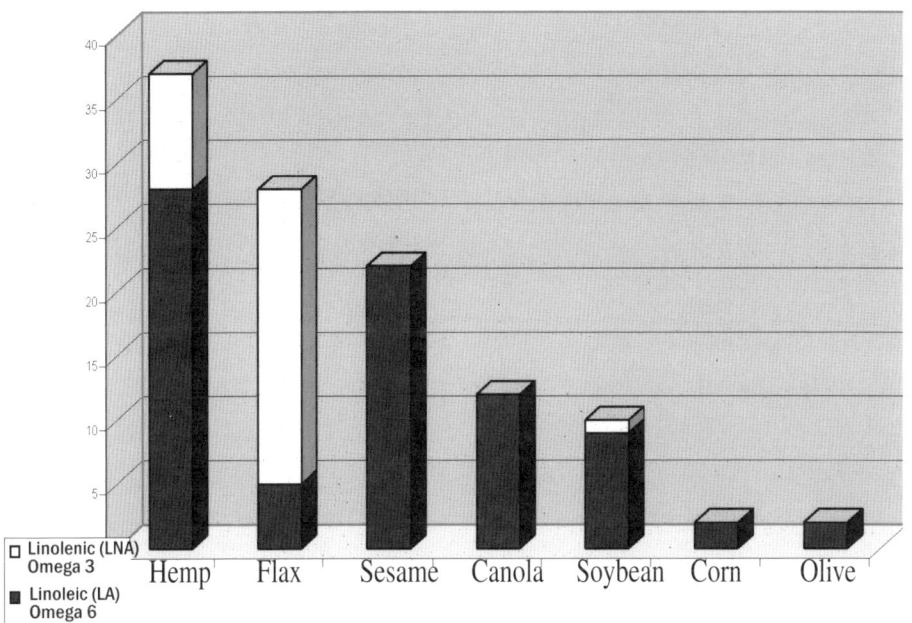

EFAs are important precursors to *eicosanoids* that are required by energy and oxygen demanding tissues such as the brain, retina, adrenal and testicular tissues.

The metabolism of all these fats occurs in the liver. Fats must be converted into substances the body needs, used as energy, or ejected as waste. Enzymes, found only in raw and mainly unprocessed foods are used to change the structure of fatty-acids.

EFAs, because they become part of the structural components of cell membranes, also play an important role in the transfer of nutrients and wastes to and from the cell. The ability of EFAs to hold oxygen in our cell membranes helps to protect cells from toxins.

EFAs and Your Baby

Omega-3 EFAs are shown to be beneficial to the growth and development of a healthy child. A large part of the baby's brain growth takes place during the

> Many women report improved nail strength, improved skin texture and appearance, lustrous shiny hair and lack of cravings for fatty junk food when supplementing their diet with omega-3 rich foods.

last three months of pregnancy. Babies born too early run the risk of not getting all the omega-3 needed to complete this process. Pregnant mothers, growing teenagers and children with behaviour difficulties will all benefit from a good intake of well-balanced EFAs. Scientists stress the value of breast-feeding as breast milk naturally contains a good supply of omega-3s. Omega-3s play a critical role in the development of brain and visual functions and may help prevent diseases such as *Candida albicans*, allergies and emotional hyperactivity disorders.

EFAs are so important to baby's growth that The Health Protection Branch of the Canadian Government, the equivalent of the American FDA, is considering requiring that all infant formulas contain adequate amounts of omega-3.

Despite all these facts, technological developments over the last 100 years have contributed significantly to a shift in our diets from the natural fats to the bad *trans*-fatty acids. This

> During breast feeding, the child pulls 11 grams of EFAs from the mother daily through breast milk.

is due mainly to the increase of processed foods in our diets, especially processed foods that contain hydrogenated fats and oils. Hydrogenated fats are often deceptively labeled on products as 'fats' and/or 'vegetable oil'.

Experts believe that despite the increase in medical technology and sanitation, this shift towards bad, hydrogenated fats has led to the reduction in life span of our civilized society due to their highly damaging effects on our immune systems. Remember to look for hydrogenated on the label of all your foods that contain vegetable oils and avoid them as you would a poison. If you are not sure if your favourite products' fats are hydrogenated then call the manufacturer.

Hemp and Soy

The hemp seed, being so high in protein, can be used as a substitute for meat or milk in much the same way as the soy bean is used.

Some studies have shown soy to not be a suitable food for young babies. Fresh hemp milk, free of Genetic Modification may be a more suitable weaning alternative.

Hemp seed does not contain the anti-nutrient trypsin inhibitors, found in most soy products.

One tablespoon of hemp oil taken daily, over a 12-week period can significantly improve PMS-related symptoms.

PLEASE CONSULT A MEDICAL DOCTOR BEFORE CHANGING YOUR BABY'S DIET

"The developing fetus literally sucks essential fats out of the mother's body because these fats are so vital for its brain development," says Dr. Erasmus. "Research suggests this depletion of the mother's essential fats is a main reason why women get postpartum depression, and why more women than men suffer from depression, lupus, multiple sclerosis, and fibromyalia."

Hemp as a Healer

The protein in hemp may relieve the symptoms of many diseases, including Parkinson's and AIDS. Hemp seed can be beneficial to those suffering from liver and kidney disorders because it contains albumin, a protein usually manufactured by the liver. Glutamic acid, a neurotransmitter, is found in good quantities in hemp seed, and may alleviate stress-related symptoms. Hemp seed has also been positively associated with treatment of symptoms caused by Alzheimer's disease and arteriosclerosis.

I have been using hemp seed for over ten years now. Addition of hemp in my diet has certainly helped eliminate my cravings for negative fatty foods. Recently, since becoming a father many have commented how I look younger than I actually am, when I expected quite the opposite! Adding hemp seed oil, nut or whole seed to my diet has affected me in immeasurable ways. Eating live hemp foods, as detailed in my Living Foods Recipes book, I have felt a difference in all areas of my life. I have more energy, think more clearly, need less sleep and find myself more productive during the time that I am awake. Physically, my skin became silky smooth pretty soon after I started ingesting hemp seed products. 10 years later, it still is. I look forward to my advanced years now, with the knowledge that my glowing complexion is likely to stay!

And it is not just me. Along my journey I have introduced many others to the benefits of hemp seed and oil. I continue to be astonished by the reported benefits. Ravi found that hemp oil added to lotions and conditioners gave a deeper moisturizing to the skin, scalp and hair. With repeated use it helped him protect from dry, flaking and cracking skin.

> "Over the last 10 years, numerous anecdotal reports have claimed that hempseed oil improves skin integrity, strengthens finger-nails and thickens hair," informs Dr. J.C. Callaway

Tanya, a 14 year old from London suffered from eczema, a dry skin condition since she was just a year old. Initially it only affected the skin behind her knees and could be controlled with emollient creams supplied by her doctor. However, when she was 11, Tanya's eczema suddenly worsened, eventually spreading over her whole body. Despite potent steroid creams and frequent bandaging, her condition did not improve. After being recommended hemp, her skin became less dry and irritated and she stopped scratching as much and was able to get by without most of her bandages. After several weeks Tanya's skin, which had been flaky and leathery, started to smooth out. This was thanks to hemp seed oil.

Recent clinical trials from Finland also indicate a remarkable reduction in dryness, itching and an overall improvement in the symptoms of the patients while they were using hemp oil.

Doris Triber, an active 84 year old from Essex, heard about the benefits of hemp seed in her local health food store. She was determined to

> When penicillin was introduced to modern medicine, it sat shelved for some time until it's true potential was fully realised.

try hemp for her health, but her teeth were not like they used to be. Hemp seed oil was not available to her at the time, but a high hemp content snack bar was. She consumed one snack bar every day, which was a feat alone for her dentures! Doris contacted me via her daughter who was astounded at the difference. "Doris has always been an active soul, but since regularly eating these snack bars she bounds with energy like she was half her age!"says her daughter, Jackie.

Doris, who has been eating hemp food products for over one year now has not had a cold or any other illness all winter. Hemp seeds' essential fatty acid content helped boost her immune system, and provided her with extra energy, whilst the quality protein and calcium content of the snack bar assisted in keeping her bones healthy.

Matthew had been diagnosed hypoglycemic, a blood sugar level imbalance, for 1   years. He had notable allergies to various foods, including all wheat products. He found himself having to eat a lot of meat to rid himself of his cravings. One day a friend suggested trying hempseed. He used hempseed every day for one full week. The meat cravings disappeared! Two months later he became so used to not having cravings any more that he left hempseed out of his diet, and the cravings came back. Now hempseed is a regular part of his diet.

Michelle, mother of 2 took hemp seed oil for the full term of her pregnancy. The essential fats from hemp seed are said to aid in the growth of the child's brain during the last 3 months of pregnancy. Michelle continues to take hemp seed oil and hemp products regularly now as the essential fats are transferred to the child through breast milk and may prevent the child from acquiring *Candida*, a wide-spread condition.

US State Representative Cynthia Thielen sponsored legislation that established the Hawaii Industrial Hemp Research Project. Hemp has the potential to replace sugar cane as a major crop, once federal restrictions are changed. Representative Thielen stated: "On a personal note, I have benefited greatly from the use of hemp oil as a treatment for sun-damaged skin. After having a cancerous basel cell removed, I started ingesting one tablespoon of hemp oil daily. My dermatologist credits the hemp oil for the rapid and cosmetically successful healing of my incision.

Attention Deficit Disorder

A recently discovered benefit of EFAs is their positive role in the treatment of Attention Deficit Disorder (ADD). Many hyperactive children show signs of EFA deficiency including severe thirst, dry skin, and atopic allergies. Such persons can benefit from supplementing their diets with foods that contain EFAs, such as hemp rather than taking prescription psychiatric drugs such as Ritalin.
The inability to metabolize essential fats correctly has also been linked to other diseases including multiple sclerosis, adult depression and dyslexia. The use of various hemp seed based products may contribute to alleviating these symptoms.

Arthritis and Osteoporosis

A daily dose of approximately three tablespoons of GLA, taken orally over a period of twelve weeks, was found to reduce the symptoms of rheumatoid

Traditional Chinese Medicine
China is the biggest producer of hemp seeds. Rooted in the history of Chinese medicine, hemp is listed among the superior elixirs of immortality that could be taken for indefinite periods of time. A recent translation, of the Pen T`sao Kang Mu, a 400-year-old Chinese medical text, describes the use of hemp seeds: "To mend and help all of the central areas of the chi. The Ancients used this medicine to remain fertile, strong and vigorous, and not be subject to aging... It has the capacity to cure neurological impairment due to stroke and the problems of excess sweating which it brings on... It improves the urinary tract and the passing of urine. It can break up long-standing problems with the blood flow. It will restore the blood, pulse, the veins and arteries. It will alleviate retained placenta illness in mothers just beginning to suckle their infants. If one's head is washed with this, the hair will accelerate its growth, and be properly balanced with just the right amount of moisture."

arthritis, without any side effects. Supplementation of GLA and vitamin D has been shown to increase calcium absorption, improve bone strength and help to prevent osteoporosis and kidney stones.

Hemp oil contains between two to four percent GLA and 100iu/100g of vitamin D. Using an oil that contains naturally occurring GLA and vitamins in conservative quantities is recommended over using a tablet form of supplementation. Recently Motherhemp Ltd, a company in the UK, announced clinical trial results showing that hemp seed oil caused a dramatic increase in blood level GLA. Increased serum levels of GLA might help explain some of the numerous anecdotal reports of seemingly miraculous cures from people taking hemp seed oil, particularly those suffering from chronic health problems such as allergies, dry skin, slow wound healing, and rheumatoid arthritis.

High Cholesterol

High levels of low-density lipoprotein (LDL), the "bad cholesterol," are a major contributor to arterial plaque, the fatty deposits on the interior walls

> Phytosterols, a component of hemp oil have been shown to reduce total serum cholesterol by an average of 10% and LDL cholesterol by an average of 13%.

of blood vessels. Over time this can lead to arteriosclerosis, a common cardiovascular disease. In a recent study, patients who increased their daily dose of LA and GLA, equivalent to five teaspoons of hemp oil per day, were shown to rapidly decrease their elevated blood levels of LDL cholesterol. They also reduced their risk of thrombosis due to the relatively high phytosterol content of hemp seed (438 milligrams of phytosterol per 100 grams of seed). Increasing intake of natural fibre, such as that found in whole hemp seed, also helps to control cholesterol.

Hypertension

With increased dietary intake of EFAs, hypertension is relieved and the risk of heart attack and stroke is reduced. The chances of succumbing to any heart related disease in general is significantly reduced with the consumption of the right balance of EFAs, and accompanying vitamins and minerals. Numerous human and animal studies have shown that substitution of non-hydrogenated polyunsaturated fats for saturated fats can reduce the risk of sudden cardiac arrest and fatal cardiac arrhythmia. In 2000, the American Heart Association issued a recommendation that Americans consume foods with higher levels of omega 3. Hemp fits the bill perfectly.

> High Omega-3 fatty acid containing foods, such as hemp have been suggested as acting as mood stabilizers in persons with depression and mania. Hemp foods may represent the future in designer mood stabilizing foods!

Immune Health

Another major benefit of hemp oil is a strengthening of the immune system. It inhibits tumor growth, kills bacteria (including *Staphylococcus*), and heals wounds.

Tuberculosis

In a Czechoslovakia during and after World War II, hemp protein was used to treat tuberculosis. At institutions for children with tuberculosis, doctors had no medicine and very little food so they decided to treat the children with hemp seeds, because of the edestin it contains. Edestin has the appropriate amino acids (including arginine, essential for formation and growth of new tissue) useful for a wealth of healthy enzymes. Twenty-six children were treated with a diet of hemp seed, oats, and cottage cheese. All were cured, or significantly improved, and grew to be healthy young adults.

"Grow more….Know more
Learn more…. Earn more
And find… Health truly is the Greatest Wealth"

Non-edible Hemp Products

What Else is Hemp Good For?

✓ TEXTILES- soft and strong. Can enhance other materials.

✓ PAPER- sustainable option to logging ancient forests

✓ FUEL- any diesel engine will run on oil derived from the hemp plant

✓ PLASTIC- renewable alternative to petro-chemical based plastics

✓ BODY CARE- a nutritious, non-toxic base for body-care products

✓ ANIMAL FEED- EFAs are beneficial to animals as well as humans

✓ HORSE BEDDING- more absorbent than straw, also used as a 'mulch'

✓ BUILDING MATERIALS- stronger and lighter than cement

✓ CAR PARTS- now used by the world's leading manufacturers

✓ MEDICINE- superior for pain and stress relief

Textiles

Hemp textiles have been manufactured for centuries. The fibres are strong, breathable and hard-wearing. A preferred option to growing cotton that is renown for its high pesticide and water requirements.

PHOTO: EIL

Modern Italian and Chinese fibre mills produce hemp materials of extraordinary quality. Recent technology allows hemp fibre to be blended with silk, lycra, fleece and other materials; modern hemp textiles are no longer the coarse material often associated with ropes and sail materials.

Paper

Hemp has a long history in the production of paper. 2000 years after it was buried, the first known piece of hemp content paper was found in a tomb in the Shensi province of China, dated 100 BC. Until the Industrial Revolution, most of the paper in the world was made from hemp and flax fibres. Renewed interest in hemp fibre for paper production is partly due to rapidly disappearing forests. Hemp can produce up to four times more useable fibre than most forests, over the same timespans. When wood pulp was first used, in the early 20th century, trees were abundant and energy costs negligible, so fibre extraction was considered very economical. This is no longer the case. As logging and energy become more expensive, fibre extraction from such hard material is less viable. By mixing hemp with other renewable fibre sources such as post-consumer waste, flax, and straw, hemp can enhance paper strength, helping it to last longer. This is yet another sustainable use for hemp.

Flax and hemp yield longer fibres and can assist in creating high quality paper when added to shorter fibre resources such as recycled office paper (post-consumer waste).

Hemp Paper Facts

Acre-for-acre, Hemp can produce up to four times more useable fibre than most forests, over the same timespans.

Hemp paper does not need chlorine bleach, which heavily pollutes rivers near wood-pulp paper mills.

Hemp paper is stronger, finer and lasts longer than wood-based papers.

Today, hemp paper is used for archival paper, art paper, tea bags and currency notes.

Perhaps the oldest specimens of paper were discovered in a tomb near Xian in Shensi province. Dated no later than 87 BC the pieces were wrapped in hemp cloth.

The pulp and paper industry is the fifth largest consumer of energy, accounting for 4 percent of all the world's energy use.

Body Care Products

Traditional Hemp Hair Growth Formula

"Boil hemp seeds until they become black. Remove the seeds and extract the oil by crushing the seeds. The remaining liquid can be used as a drink to soothe the throat and stop coughing. The extracted oil, ma you, is applied to the scalp to clean the pores and feed the hair roots. Used to cure excessive hair loss and stimulate overall hair growth."
—From the sixteenth-century Pen T'sao Kang Mu, translated by Norman Goundry

Experiments have shown that the hemp oil, due to its high EFA content, rehydrating and anti-inflammatory properties, is better for skin and hair than vegetable oils such as avocado, almond, or olive oil. Hemp oil is now used as a base in soaps, creams, shampoos, shower creams, lip balms, bathing essences, and ointments. EFAs work from the inside and out to increase the smoothness of skin and the health of hair. EFAs are used by the body to build and maintain healthy body cells (especially cell membranes), and work directly on epidermal cells, entering the lipid layers of dry skin cells to replenish their oils. EFAs also repair skin damage, help heal wounds and burns, and have antibiotic properties. Research has shown that EFAs are an effective treatment for atopic dermatitis, eczema and psoriasis.

SEE RESOURCE SECTION PAGE 49 FOR SUPPLIERS OF QUALITY BODY CARE PRODUCTS

Although body care products are manufactured using refined hemp oil, which is less susceptible to deterioration than food-grade oil, hemp-based body care products must still be protected from heat, oxygen, and light. Light is particularly damaging, which is why hemp products usually come in opaque containers, and is why hemp should never be used on the skin as a sunscreen.

Cleaning Products

Hemp has great potential for use in detergents and cleaning products. Some German companies have developed products that use hemp oil, rather than petrochemicals, in a range of detergents. Hemp detergents are fully biodegradable and less of a burden on the environment.

Bio-Fuel

Bio-diesel is a fuel made from any vegetable oil and can be used to fuel any diesel engine. The most commonly used oil for fuel production is canola, or rapeseed oil, though hemp works well. The by-products of bio-diesel processing are oil-seed cake (a high protein animal feed), and glycerol (which is used in soap making).

"Why use up the forests which were centuries in the making and the mines which required ages to lay down if we can get the equivalent of forests and mineral products from the annual growth of the fields?" —Henry Ford

Henry Ford demonstrating the strength of 'cars made from plants' – the axe bounced back off the car, without a dent"

Bio-diesel has virtually identical power output and fuel economy to fossil diesel. In addition, it produces less smoke, virtually no sulfur dioxide emissions and significantly reduced carbon monoxide emissions. The consequences of bio-diesel fuel leaks or spills are far less concerning than with petro-chemical diesel leaks -

98 percent of bio-diesel breaks down within 21 days.

A. Das of Original Sources was the first to make biodiesel from pure hemp oil. My friend D. Maxwell from New Earth, Montreal, has also powered his car on bio-diesel for more than ten years. Hemp oil powered cars are touring the world, from North America to Japan. These cars are showing that we can 'grow our own fuel' rather than rely on Middle Eastern oil supplies. Recent tragedies have shown such reliances on petro-chemical oil as having unwanted consequences.

> Biodiesel has virtually identical power output and fuel economy to fossil diesel. In addition, biodiesel produces less smoke, virtually no sulphur dioxide emissions and significantly reduces carbon monoxide emissions.

Other Commercial Hemp Products

Non-woven hemp and flax fibres are used for production of plant pot liners and mulch mats. Hemp fibre is used as a base for particle-boards that are competitively priced alternatives to conventional wood products. Hemp is also used to make building materials and insulation, carpet backing, animal bedding and moulded products as diverse as pots, credit cards, car parts, furniture and toys.

> Widespread use of hemp plastics may reduce the consumption of polluting petrochemical based plastics

Hemp Plastic

Hemp now promises to play its part in the plastics industry as an alternative to non-sustainable petrochemicals. Various molded products using a number of techniques can be produced from environmentally controlled biodegradeable materials. Hemp seems to bring us many of the solutions required in the 21st century.

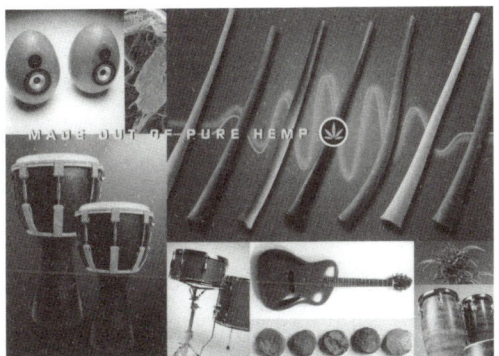

The didgeridoos and other instruments pictured left, are the first commercial products made from **100% hemp**- a material known as 'Hemp Stone'. They are as hard as stone with acoustic qualities said to be world class. The indigenous people of Australia have become involved in this project and are supporting the widespread growth of hemp as they recognise its sustainability. These didgeridoos have been painted by Aboriginal artists and production facilities are expanding now.

Hemp Plastic has been shown to be useful in production of shelving, cd-cases, credit cards, car parts, furniture, musical instruments, bowls, cosmetic containers, shopping bags, surf-boards and children's toys!
The hemp frisbee 'High Fly' (1998) and the CD case of 'Fields of Green' (2003) were the first commercial hemp plastic products since Henry Ford first introduced the idea.

THE FIRST HEMP CD PRODUCED ON HEMP PLASTIC CD TRAY, USING HEMP PAPER AND SITTING ON A FALLEN TREE NOT NEEDED FOR ANY OF THESE MATERIALS!

HAVE FUN WITH A 25% HEMP PLASTIC FRISBEE.

Pharmaceuticals

The drug producing cousin of hemp, commonly known as marijuana or Medicinal Cannabis usually contains around 2-20% of the psychoactive ingredient THC. Medicinal cannabis is derived from the flowering tops of high THC producing *Cannabis* plants.

In Britain, cannabis was medically available until 1971. In 2004, trials led by GW Pharmaceuticals show medicinal cannabis is soon to become available on general prescription for various ailments.

Medicinal cannabis is said to be useful for, but not limited to: AIDS Wasting Syndrome, Anti-Tumor Effects, Arthritis, Asthma, Brain Injury/ Stroke, Crohn's Disease and Ulcerative Colitis, Depression and Mental Illness, Eating Disorders, Epilepsy, Fibromyalgia, Glaucoma, High Blood Pressure/ Hypertension, Migraine, Multiple Sclerosis, Nail Patella Syndrome, Nausea association with Cancer Chemotherapy, Pain reduction and tolerance of Pain, Phantom Limb Pain, Schizophrenia, Spinal Cord Injury and Tourette's Syndrome. Whew!

Medicinal Cannabis is most frequently administered via an inhaler, under the tongue spray or tablet

High-THC cannabis can produce side effects including drowsiness and lack of co-ordination. However, when compared to the toxic side effects of alcohol or tobacco, cannabis is considered more acceptable. Cannabis users are often relaxed rather than rowdy, and despite millions of deaths directly related to alcohol and tobacco abuse, there has not been a single occurrence of death due to the use of just cannabis, anywhere in the world.

The Hebrews have had a long and beneficial relationship with this useful plant. It was known as qaneh-bosm, a possible source for the root name *Cannabis*.

The difference between recreational use and medicinal use of cannabis is mostly in the intention of the user

Glossary

Essential Fatty-Acids (EFAs)— two fatty-acids, linoleic acid (LA) and alpha-linolenic acid(LNA), that cannot be made in the body and must be acquired from food. Hemp seed is the best naturally balanced source of these EFAs. Both are sensitive to light, oxygen, and heat. Absence of LA or LNA is fatal. LNA is less abundant in modern western diets than 100 years ago.

Hemp Flour—coarse or finely ground hemp seed. Can be bought or freshly ground in a coffee grinder or hand mill.

Hemp Milk hemp seeds, water, sweetener, and flavour blended and strained. May be used as a beverage or dairy substitute.

Hemp Nut—sometimes referred to as soft hemp seed or dehulled hemp seed, it is the "meat" of the whole, raw hemp seed. A mechanical process removes the crunchy hull, leaving the soft white nutrition-packed interior.

Hemp Hull—residue from production of hemp nut via dehulling. Used as a high fibre animal feed, a cushion fill or mulch.

Hemp Oil—Hemp oil may be extracted by mechanically or chemically pressing hemp seeds. The optimum method of pressing oil is without light and oxygen, and with as little incidental heat as possible. This helps slow down free-radical production, which can be detrimental to health. Refined hemp oil is usually clear in color and used as a base for body care products. Unrefined hemp oil is usually green in color and is best as a food source.

Hemp Seed—the whole nut of the *Cannabis* plant that contains only traces of THC.

Should hemp seed be hard to source in your area you may substitute:
Hemp Milk <> Almond milk
Hemp Seeds/ Nut<> Sunflower seeds and/ or flax seeds
Hemp Butter <> Cashew, almond butter or tahini. Keep refridgerated.
Hemp Oil <> Flax Oil/ Udo's Choice™ . Keep refrigerated
Hemp Flour <> Hemp cake or LSA (freshly ground flax seed meal). Keep refrigerated

DEHULLED HEMP SEED

Various parts of the hemp seed go together to produce a high protein, fibre and omega-3, 6 and 9 source in various foods sold throughout the world, being most popular in Europe and North America

WHOL HEMI SEED

HEMP HARVESTING IN NORTH WALES, UK

HEMP-ICE
DAIRY FREE- ORGANIC-
HEALTHY AND DELICIOUS!!

HEMP SEED ADDS
NUTRITION TO
SUNNYVALE BRAND
SPROUTED WHEAT BREAD

PHOTO: HEMPCAR.ORG
THIS MERCEDES DROVE THE ENTIRE
LENGTH OF THE USA ON HEMP BIO-DIESEL

PHOTO: MOTHERHEMI

HEMP HARVEST, TASMANIA, AUSTRALIA

PHOTO: EIL

HEMP AS AN ALTERNATIVE TO MDF BOARDS

PHOTO: DAIMLER CHRYSLER

PHOTO: EIL

HEMP HURDS BEING SOLD AS AGRISORB™ GARDEN MULCH AT A MAINSTREAM HARDWARE STORE

PHOTO: HEMPSTONE

INSTRUMENTS MADE FROM 100% HEMP STONE

THE AUTOMOTIVE INDUSTRY HAS FOUND BENEFITS FOR HEMP IN VARIOUS BODY PARTS INCLUDING INTERIOR LININGS, SPOILERS AND BREAK LININGS. QUALITIES OF HEMP INCLUDE STRENGTH, WEIGHT, DURABILITY AND COST EFFECTIVENESS

HEMP GEOTEXTILES FOR MULCH MATS AND MORE FROM INDUSTRIAL HEMP

A FARMER ON ANGLESEY, UK FINDS OUT JUST HOW TOUGH THOSE HEMP FIBRES CAN BE

Resources

Food & Oil

Hemp Food Industries Association (HFIA) is the global information source for all things hemp. The HFIA can connect you with a local manufacturer, wholesaler or re-seller. You can support the HFIA's mission by joining as a member- more details on their web site.

For more information on Fats and Oils visit www.udoerasmus.com

Fibre & Plastics

Various organisations can assist in growing & processing hemp.

UK, Canada and Australia are currently the most popular places to grow hemp and licenses are allowed in all of these countries. Many new and improved facilities and systems are opening up around the world. The HFIA is your connection to up-to-date information.

Medicine

For corporate medicinal cannabis products visit www.gwpharm.com who are pioneering the way forward.

Flowering tops and by-products have long been sought after globally. There are currently only some countries that allow public access to these varieties of hemp.

Your one stop contact for all things hemp:

www.hemp.co.uk

HFIA- we are here to help

A MODERN INTRODUCTION TO HEMP

The Hash Marihuana & Hemp Museum

A M S T E R D A M

A PERMANENT EXHIBITION ABOUT THE HISTORY
AND USE OF CANNABIS AROUND THE WORLD.

*No trip to Amsterdam is complete without
a visit to this unique museum.*

*Learn about the amazing history of the
Cannabis plant and see a live indoor
Cannabis garden.*

The Hash Marihuana & Hemp Museum

Oudezijds Achterburgwal 130 Amsterdam. Next to the 'Sensi Seed Bank' Grow Shop.
Open all week from 11.00 until 22.00 hours. www.hashmuseum.com

ABOUT THE AUTHOR

Paul Benhaim is published author of H.EM.P., Healthy Eating Made Possible and Living Food Recipes

In 1991 he followed his heart and produced the first commercially available hemp snack bar in the EU, pioneering the hemp food industry. His first product is now sold in mainstream supermarkets world-wide.

Currently Managing Director for Hemp Foods Australia where he has brought his skills in pioneering improved recipes, formulas and production capabilities to bring hemp foods to an even wider audience.

Creating his first hemp plastic product in 1998, Paul is now offering mass produced hemp plastic products. His latest hemp plastic product- a CD-tray was introduced to market by using a relaxation music CD that he designed, produced and played a hemp didgeridoo for.

Manager of some of the world's most popular hemp web sites and advisor to a number of companies world-wide, he is also involved with writing and lecturing on Living Foods.

A qualified yoga teacher, masseur and a keen gardener, Paul is often found walking near his rainforest home of Northern NSW, Australia.